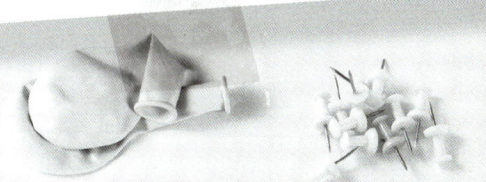

SCIENCE MAKERS

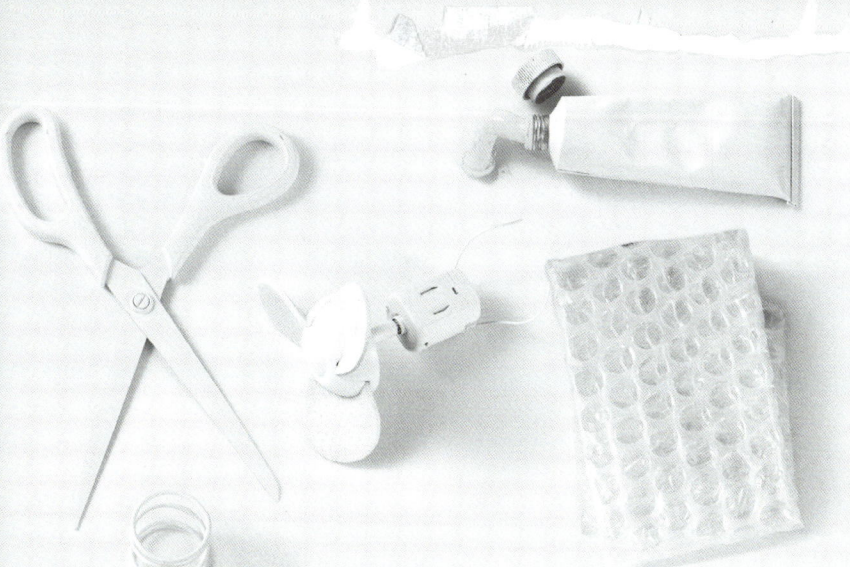

Making with
SOUND

Anna Claybourne

BUILD **AMAZING PROJECTS** WITH INSPIRATIONAL SCIENTISTS, AR̲_____S

Published in paperback in Great Britain in 2019 by Wayland
Copyright © Hodder and Stoughton, 2018
All rights reserved.

Editor: Sarah Silver
Designer: Eoin Norton
Picture researcher: Diana Morris

ISBN 978 1 5263 0547 3

All photographs by Eoin Norton for Wayland except the following:
Cornelis Bloemaert 1665. Germanisches Nationalmuseum/PD/CC Wikimedia Commons:
14tl. Blue Ring Media/Dreamstime: 5t. Goran Bogicevic/Shutterstock: 5bc. Matthew Brady
1878. LOC/CC Wikimedia Commons: 20t. Gilbert H. Grosvenor Collection, Library of
Congress/PD/Wikimedia Commons: 5c. Hulton Archive/Getty Images: 10c. Courtesy of
the Jones River Village Historical Society and the Kingston Public Library Local History
Room: 26tl. kndynt2099. cc-by-sa-2.0/CC Wikimedia Commons: 12t. Igor Kruglikov/
Shutterstock: 17b. La Gorda/Shutterstock: 5bl & r. © Kym Maxwell: 22t. Nerjon Photo/
Shutterstock: 23br. office museum/PD/Wikimedia Commons: 8cr. Fabio Da Paola/REX/
Shutterstock: 6tl. Ana Aguirre Perez/Shutterstock: 11b. Public Domain/Wikimedia Commons:
8tl. rawcaptured photography/Shutterstock: 21br. Asier Romero/Shutterstock: 15b. shu2260/
Shutterstock: 4t. Dmitry Skvortsov/Shutterstock: 5tl. Staats Sächsische Landesbibliothek
und Universitåtbibliothek Dresden/PD CC Wikimedia Commons: 14cr. Stock-Vector Sale/
Shutterstock: 29br. University Archives, Kenneth Spencer Research Library, University of
Kansas Libraries: 16tl. U.S. Navy photo by Photographer's Mate Airman Matthew Clayborne/
Wikimedia Commons: 9b. US Patent Office/PD/Wikimedia Commons: 16bl, 26tr. M Watts-
Hughes. The Eidophone Voice Figures, 1904: 18tl. 18tr, 19br. Wollertz/Shutterstock: 28tr.

Every attempt has been made to clear copyright. Should there be any
inadvertent omission please apply to the publisher for rectification.

Printed in Dubai

Wayland, an imprint of
Hachette Children's Group
Part of Hodder and Stoughton
Carmelite House
50 Victoria Embankment
London EC4Y 0DZ
An Hachette UK Company
www.hachette.co.uk
www.hachettechildrens.co.uk

Note:
In preparation of this book, all due care has been exercised with regard to the instructions, activities
and techniques depicted. The publishers regret that they can accept no liability for any loss or injury
sustained. Always follow manufacturers' advice when using electric and battery-powered appliances.

The website addresses (URLs) included in this book were valid at the time of going to press. However, because
of the nature of the Internet, it is possible that some addresses may have changed, or sites may have changed
or closed down since publication. While the author and publishers regret any inconvenience this may cause to
the readers, no responsibility for any such changes can be accepted by either the author or the publishers.

CONTENTS

TAKE CARE!

These projects can be made with everyday objects, materials and tools that you can find at home, or in a supermarket, hobby store or DIY store. However, some do involve working with things that are sharp or breakable, or need extra strength to operate. Make sure you have an adult on hand to supervise and to help with anything that could be dangerous, and get permission before you try out any of the projects.

UNDERSTANDING SOUND

Whatever you're doing, you're almost always surrounded by sound – music playing, people talking, traffic noise, alarms pinging or birds singing. It's a big part of most people's lives. Humans (like many animals) use sound to communicate, in the form of speech, sirens and alarms. And music, singing and dancing are a huge part of our culture. Many inventions involve sound, from the first musical instruments to modern smartphones. Sound is used in the art world too, in sculptures and installations that make sounds for the audience to experience.

WHAT IS SOUND?

Sound is a form of energy, like light, heat, electricity, or movement. In fact, sound is made by objects or materials moving in a particular way – shaking quickly to and fro, or vibrating.

As the material vibrates, it pushes against the molecules in the air around it, making them vibrate too. Pressure waves of vibrating molecules, called sound waves, then spread out through the air. We hear sounds because our ears can detect the sound waves, and convert them into signals to send to the brain. The brain then interprets and understands them.

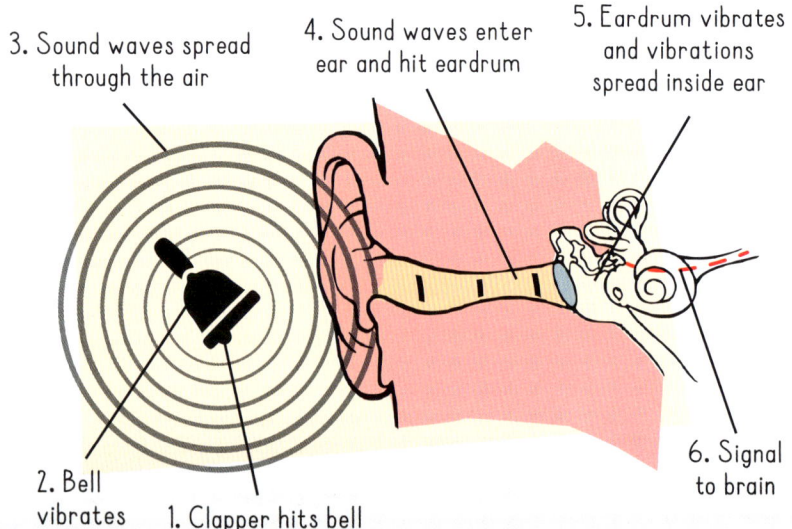

3. Sound waves spread through the air

4. Sound waves enter ear and hit eardrum

5. Eardrum vibrates and vibrations spread inside ear

6. Signal to brain

2. Bell vibrates

1. Clapper hits bell

THAT'S A BELL RINGING ...

All the different sounds we can experience, with all their different qualities, are simply different patterns of vibrations. The speed of the vibrations decides the pitch – how high or low a sound is. The strength of the vibrations – how much each molecule vibrates to and fro – decides the volume of a sound. Sound waves can also travel through liquids and solids, so sea creatures can hear underwater, for example.

GOOD VIBRATIONS

When you hear lots of different noises at once, such as the sound of a busy café or an orchestra playing, your ears are receiving a complex mixture of different vibrations, all mixed together. We can also 'feel' some sounds with other parts of our bodies, such as feeling vibrations through the ground at a loud rock concert. Some animals also sense sound vibrations through the ground.

Dolphins make clicking sounds underwater, and listen to the echoes that bounce back off objects, helping them to navigate and find food.

Inventor Alexander Graham Bell tests an early telephone line, linking New York to Chicago, in 1892.

SOUND INVENTIONS

Sound is central to many of the inventions that feature in our everyday lives – telephones, TVs, computer games, alarm clocks and film soundtracks. The discovery of how to record and play back sounds, in the 1870s, revolutionised modern life. Sound is used in other inventions too, such as ultrasound scanners that can look inside the body, or sonar, which (like a dolphin) uses reflected sounds to detect objects underwater.

SONIC ART

Sonic means to do with sound, and sonic art is art that makes sound. Artists have experimented with machines that make sound for over a hundred years. Strange sounds, often created using computers, are used as sound effects for computer games and films.

FOUND SOUNDS

Invent your own percussion instrument using found objects, like musician and maker Dame Evelyn Glennie.

MAKER PROFILE:

Dame Evelyn Glennie
(1965–)

Scottish musician Evelyn Glennie is one of the world's leading percussionists. She plays hundreds of different percussion instruments, including several that she has invented herself. She often uses everyday or found objects, like plumbing pipes, a car exhaust or old farm machinery — and gives her new instruments names, such as the batonka, the barimbulum and the simtak. Evelyn has been profoundly deaf since the age of 12.

WHAT YOU NEED

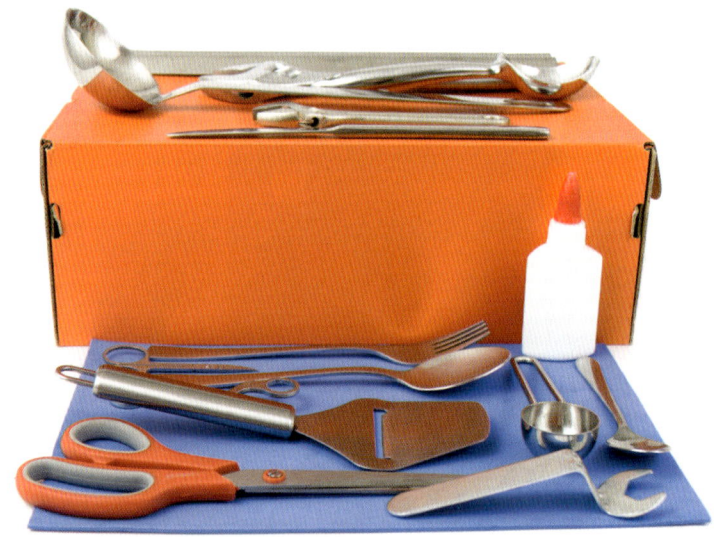

- a sturdy wooden or cardboard box, such as a shoebox
- scissors
- thick felt or craft foam
- glue
- found objects from around the home, especially metal utensils and tools, such as: knives, forks and spoons, metal bottle opener, spanners, bolts and screws, scissors, small pipes or tubes

Anything you strike, anything you shake or rattle, or just anything that can be picked up, and you can create a sound.
– Evelyn Glennie

1.

Step 1.

Cut strips of felt or foam about 1 cm wide, and slightly longer than the length of the box. If your felt or foam isn't long enough, join strips together to make them the right length.

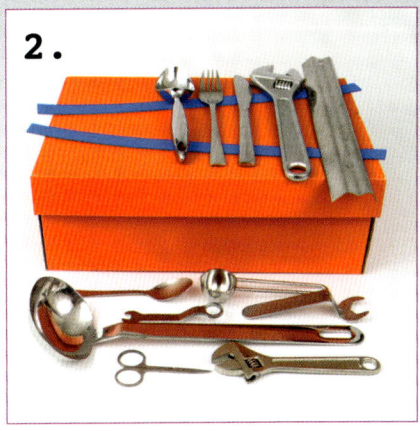

2.

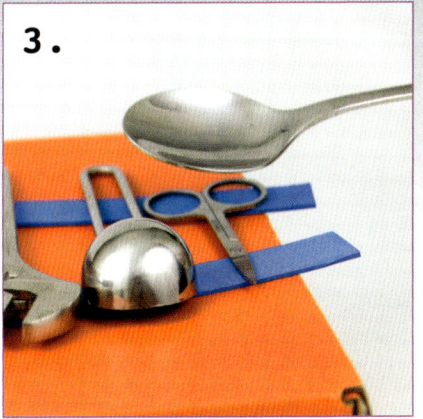

3.

Step 2.
Arrange the felt or foam strips on top of the box in a rough V shape. Lie each object across the strips, with shorter objects at the narrower end. Move the strips if necessary so the objects don't touch the box.

Step 3.
Gently tap each object with another metal object, such as a spoon, and listen to the notes each one makes. Arrange them with the lowest note at one end and the highest at the other.

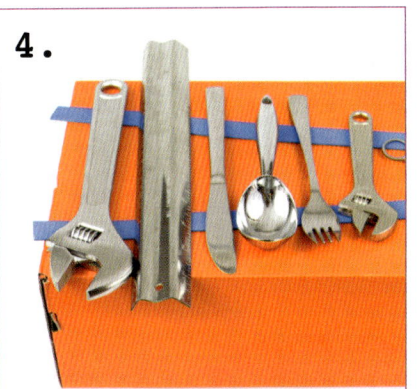

4.

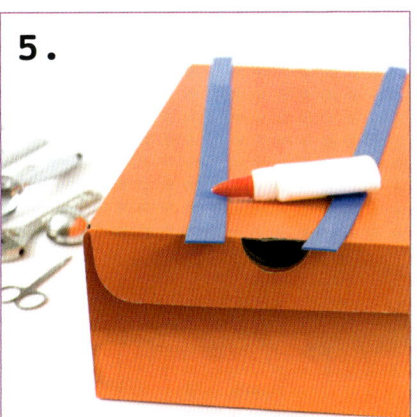

5.

Step 4.
If you don't like an object's sound, you can leave it out. Or try moving it around slightly or hitting it in a different place.

Step 5.
Once you're happy with your instrument, take the objects off (keeping them in the right order), glue the felt or foam strips in place, and put the objects back on.

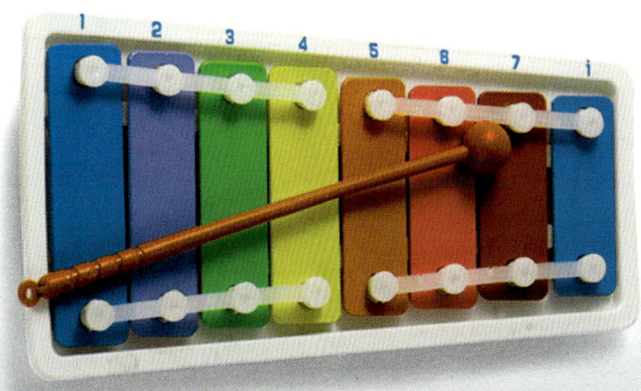

6.

Step 6.
You can play your instrument using one or two spoons. You can give it a name, too.

Even though Evelyn Glennie is deaf, she feels the vibrations from percussion instruments through other parts of her body, and 'hears' in that way. She often plays barefoot to help her feet detect vibrations through the ground.

BANGING AND SHAKING

Percussion instruments involve hitting an object. This makes the object move and vibrate in a particular way, depending on its shape and what it is made of – making that object's own distinctive sound.

Faster vibrations make a higher-pitched note, and slower vibrations make a lower-pitched note. Usually, the smaller an object is, the faster it vibrates. Using objects of different sizes lets you make a percussion instrument that can play different notes.

TALK TO THE TUBE

Link two different rooms in your home, or even two different floors, with speaking tubes!

MAKER PROFILE:

Jean-Baptiste Biot
(1774–1862)

Jean-Baptiste Biot was a French maths professor and space scientist, as well as a balloonist, inventor and all-round experimenter. In the early 1800s, he tried using long tubes to carry sounds. He found that even a REALLY long tube (almost 1 km long) could carry sounds from one end to the other, without them changing much.

Long before telephones, this discovery was used to make speaking tubes. They were used on ships to link different areas, and in large houses for people to talk to their servants in other rooms.

This office worker, at work in 1903, has four speaking tubes to use at the end of his desk.

WHAT YOU NEED

- 5 m (or more) of corrugated plastic tubing, about 3 cm in diameter
- two large plastic drinks bottles
- scissors
- strong packing tape or duct tape
- string or cable ties (optional)

Step 1.

With an adult to help, cut the tops off both the bottles, at the point where the sides of the bottle start to straighten out.

Step 2.

Wrap strong tape around the cut edges of the bottles, to cover up any sharp bits and make them more comfortable to use.

Step 3.

Attach the openings of the two bottles to the ends of the tube. You may be able to simply push or twist them together. If not, wrap tape around them to make a strong join.

Step 4.

Arrange the tube so that the ends are in two different rooms or parts of a house. You can then talk to another person by speaking into the tube while they listen at the other end. Even whispering is easy to hear!

You could use speaking tubes to link two bedrooms, or to connect the downstairs of a house to the upstairs, so that you and your parents can chat to each other without having to trek up and down. Ask an adult to help you to fix the tube in place by attaching it to bannisters or other objects, using string or cable ties.

Speaking tubes are still used today on some boats, like this US Navy ship.

DOWN THE TUBE

When you speak into a tube, your voice makes the air inside it vibrate, just like when you speak normally. But because of the tube, the vibrations don't spread out as they would normally. Instead, they are directed along inside the tube, and can travel much further. The tube itself also picks up the vibrations and carries them along. The bottle-end speakers help to collect the sound waves and funnel them into the tube, and also to make the speech sound louder when you listen at the other end.

IT CAME FROM OUTER SPACE!

If you've ever wondered how sci-fi sound effects are made – try this!

MAKER PROFILE:

Delia Derbyshire
(1937–2001)

Delia Derbyshire was a British musician and composer. In the 1960s and 1970s, she worked at the BBC Radiophonic Workshop, which was set up to make music and sound effects for TV and radio shows. Derbyshire would record voices, animal noises and sounds made by all kinds of objects – like an old metal lampshade or a glass bottle. Then she used early electronic equipment to change and shape the sounds. She's most famous for creating the sounds used in the theme music for the TV show *Dr Who*.

Sound effects for radio and TV were created using a variety of equipment at the BBC Radiophonic Workshop in London, UK.

Any sound can be made into a radiophonic sound ... The sort of sounds we usually use are electronic sounds of various sorts, and also sounds that are recorded, picked up by a microphone, everyday sounds and also musical instruments.
– Delia Derbyshire

WHAT YOU NEED

- a large metal spring toy
- two paper cups
- strong packing tape or duct tape
- pointy scissors
- spoons, chopsticks or other objects for hitting the spring
- two people

1.

2.

Step 1.

Turn a paper cup upside down so that you can see the rim around the base. Use the point of a scissor blade to make two holes in the rim, right next to the base of the cup. Do the same with the other cup.

Step 2.

Take one end of the spring and thread it through the holes in one of the cups, so that the metal lies flat against the base. Tape the spring firmly to the base. Attach the other cup to the other end of the spring in the same way, resting it on a table or fixed surface.

3 & 4.

Step 3.

Hold one of the cups, while another person holds the other one, and pull the cups away from each other so that the spring is slightly stretched out. Now put your ear to the cup, and gently tap the spring with a spoon or other object. *Pneeeoooww!*

Step 4.

Try using the device to make your voice sound alien and spooky too. Speak or sing into your cup, while the other person puts their cup to their ear. Then try it the other way round!

CHANGING VIBRATIONS

As sound vibrations move through the stretchy, bendy spring, they get distorted and changed. This ends up making many different frequencies, or vibration speeds, which make different pitches of sound. They travel along the spring and into the cups, which direct them into your ears. The sounds you hear are completely unlike normal voices or instruments, making them seem very otherworldly and 'alien'.

If you have an audio editing app, such as WavePad, Audacity or Hokusai, you could record the sounds and try making more changes to them, as Delia Derbyshire did.

AKB48 BOTTLE TRAIN

Make an automatic musical instrument that's played by a speeding train.

MAKER PROFILE:

AKB48
(Formed in 2005)

AKB48 is a huge singing and dancing girl group from Japan. There are up to 120 members at any one time, who play concerts in different smaller groups or 'teams'. The members are well-known stars in Japan. In 2016, they released a video showing a music project in which a model train played the famous *William Tell Overture* by zooming past rows of carefully arranged bottles, hitting them with drumsticks as it went by.

An AKB48 team singing and performing on stage.

WHAT YOU NEED

- at least 10 empty glass bottles, as similar as possible in size and shape
- a toy train and tracks with sloping sections
- a chopstick or long lolly stick
- two small metal bolts
- elastic bands
- sticky tape
- a jug of water
- a metal spoon

1.

Step 1.
First set up your train tracks to make a straight track that runs downhill. Put it on a hard floor with plenty of space around it.

Step 2.
Use sticky tape to attach one bolt to each end of the stick. One bolt will hit the bottles, while the other acts as a balance.

2.

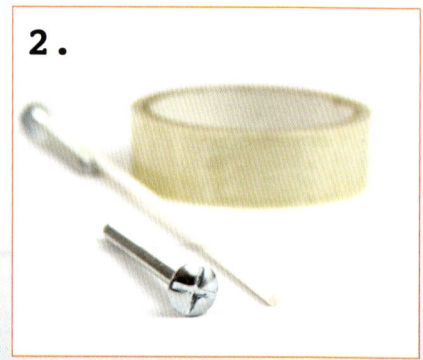

3.

Step 3.

Fix the stick to the top of the toy train using elastic bands (making sure they don't touch the wheels). The stick should be at an angle.

4.

Step 4.

Now add the bottles. Arrange them along one side of the track, and position them so that the backwards-pointing end of the stick will hit them as it passes by.

5.

Step 5.

Pour different amounts of water into the bottles to give them different pitches (notes) when hit. Test them by hitting the bottles with a spoon. Arrange the bottles so that they play a simple tune when struck in order.

Step 6.

To make the notes sound close together, stand the bottles close together. To leave longer gaps between notes, move them further apart.

Step 7.

To play the tune, put the train at the start of the track, and set it going. You may need to adjust the bottles to make it work perfectly.

6 & 7.

If you don't have a train and tracks, you could use a toy car or other vehicle. Make a simple track for it using a long strip of card with the sides folded up.

If you have a battery-powered train, you could use this instead, on a length of flat track, or a continuous circle.

SELF-PLAYING INSTRUMENTS

Automatic instruments, which play a tune by themselves, have been around since ancient times. For example, ancient Greek inventor Ctesibius made a water-powered whistling owl around 250 BCE.

The bottle train works by moving at a steady speed to hit a series of bottles. The more water a bottle contains, the lower its pitch will be. This is because water slows down the speed the glass vibrates at. More water means a slower vibration speed, and a lower note.

This wind-up musical box is an automatic instrument.

MAKER PROFILE:

Athanasius Kircher
(1602–1680)

Athanasius Kircher was a brilliant German polymath, (someone who's an expert in all kinds of different things). He studied medicine, maths, religion, earth science, ancient cultures and languages, and physics. He was very interested in sound, and invented various musical instruments. And, long before electrical amplifiers were dreamed of, he explored ways of amplifying sounds (making them louder) using horn-shaped speakers.

Using the same principle, you can make a simple smartphone speaker for playing music or radio, which uses no electricity at all.

TURN IT UP!

Use the flared shape of paper cups to make your own working phone speakers.

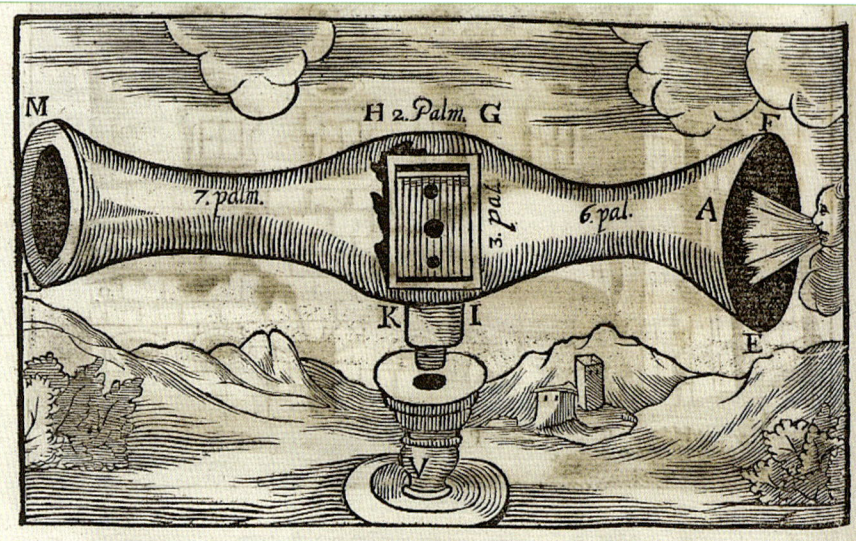

One of Kircher's sound-carrying creations.

WHAT YOU NEED

- a smartphone
- a cardboard kitchen roll tube
- two paper or plastic cups (the bigger the better)
- scissors
- a marker pen
- two paper tissues or napkins

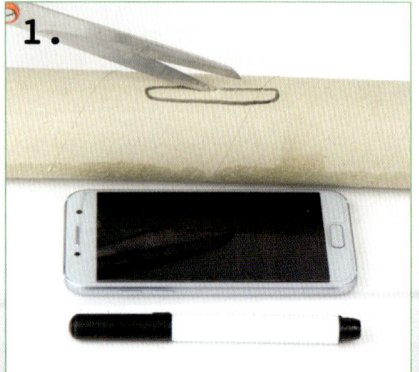

Step 1.
Hold the bottom of the phone lengthways along the cardboard tube, right in the middle. Draw around the phone with the marker pen. With an adult to help, cut out the shape of the smartphone just inside the line, using the scissors.

Step 2.
Hold one end of the tube against the side of one of the cups, close to its base. Draw around the tube onto the cup.

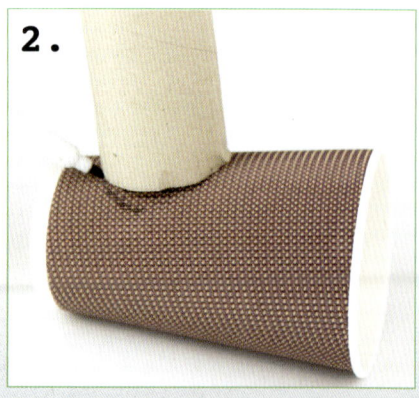

3.

Step 3.

Use the scissors (with an adult to help) to cut out the circle, again just inside the line, to ensure a close fit. Now do the same thing again with the other cup.

4.

Step 4.

Gently crumple up the two paper tissues. If they have several layers then separate to just use one layer. Push one into each end of the tube. This helps to soak up very high-pitched sounds, making the speakers less 'tinny' sounding.

5.

Step 5.

Fit the holes in the two cups onto the ends of the tube, so that both cups point in the same direction. Twist the tube so that the phone-shaped hole is pointing upwards.

6.

Step 6.

Now switch the smartphone on and play some music. Push the phone down into the hole in the tube, and the speakers will amplify it!

PASSING IT ON

The paper cup speakers work because the sound vibrations from the phone speaker pass into the cardboard tube, then into the paper cups. As the cups vibrate, they make the air inside them vibrate, and send sound waves out towards the listener. The music sounds louder because the cups focus the sound waves and point them mainly in one direction, instead of letting them spread out all over the room.

You often see horn-shaped speakers on devices for transmitting sound.

UNDERWATER EAR

Is anyone there?

It can be hard to hear what's going on underwater – unless you have a hydrophone!

MAKER PROFILE:

Lucien I. Blake
(1853–1916)

American engineer Lucien I. Blake was one of the first people to develop the hydrophone – a device for listening to underwater sounds – in the 1880s. He became interested in this after noticing that he could hear sounds from a long distance away while swimming underwater in a lake.

At first, hydrophones were designed to help ships detect signals sent out by lighthouses and other ships, to avoid collisions. But people also realised they could use them to hear the sounds of sea creatures, such as whales. Biologists still use hydrophones to listen to animal sounds in the sea.

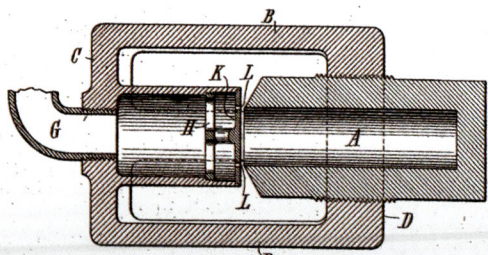

One of Blake's patents for an underwater listening device.

WHAT YOU NEED

- a length of flexible plastic tubing, about 1 m long and 2 cm in diameter
- two funnels
- a balloon
- scissors
- strong packing tape or duct tape
- a bathtub or paddling pool
- clockwork bath toys, jugs, spoons or other noisy waterproof objects

1.

Step 1.

First blow up your balloon to stretch it, and let it down again. Lie the balloon flat, and cut off the opening, just above the neck.

2.

This is tricky with a very big funnel, so if you have two of different sizes, choose the smaller one. You may need an adult to help.

3.

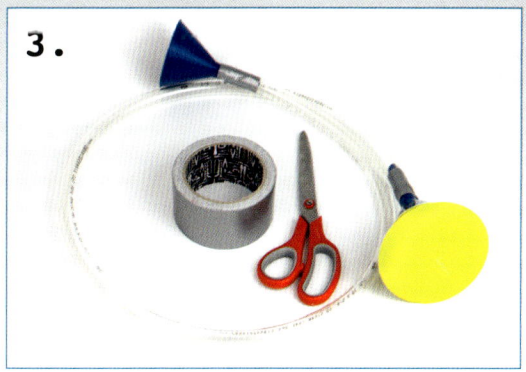

Step 2.
Stretch the remaining part of the balloon as tightly as possible over the wider opening of one of the funnels.

Step 3.
Push the two funnels into the two ends of the tube. If they don't fit tightly, use strong tape to join the funnel and the tube together.

4.

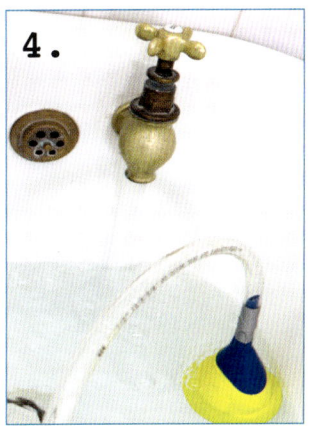

! Take care around water and always have an adult nearby.

5 & 6.

Step 4.
Half-fill a bathtub or paddling pool with water. Put the end of the tube with the balloon-covered funnel under the water surface, and listen at the other end.

Step 5.
Ask someone to make noises in the water, for example by banging two spoons together, pouring water from a jug, or putting a clockwork sea creature in the water.

Step 6.
Try listening using just your ears, and using the hydrophone. Can you hear a difference?

SOUND WAVES IN WATER

Humans use their ears to pick up sound waves that spread out through the air. Air is made of gas, but sound waves can travel through liquids and solids, too. In fact, sound waves travel faster and more efficiently through water than through air, as water is more dense and its molecules are more tightly packed together.

Humpback whales communicate by singing songs that can carry huge distances.

As our ears are built to hear in air, they don't hear very clearly underwater. But a hydrophone can collect sound wave vibrations from water, and transmit them into air, through a vibrating membrane (the tightly stretched balloon). It's then easier to hear the underwater sounds.

SEEING SOUNDS

Build a simple machine that turns your singing voice into beautiful patterns!

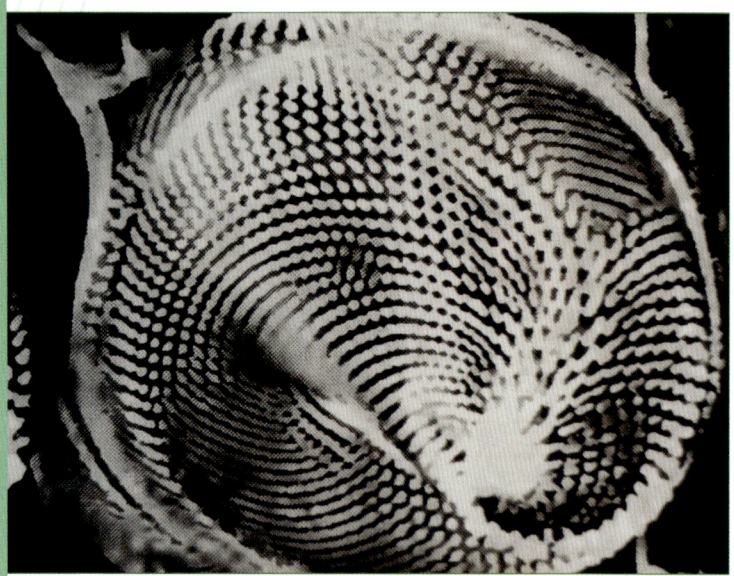

One of the patterns Megan Watts Hughes made using her eidophone.

WHAT YOU NEED

- a clean, empty cylinder-shaped cardboard container, with a pop-off plastic lid
- scissors
- a craft knife
- a pencil
- a balloon
- a small funnel
- a plastic tube about 2 cm in diameter
- strong tape
- fine glitter

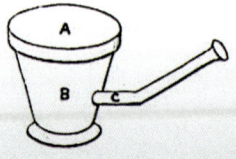
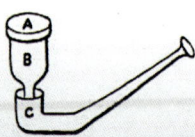

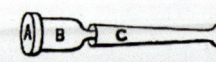

Different versions of the eidophone.

1.

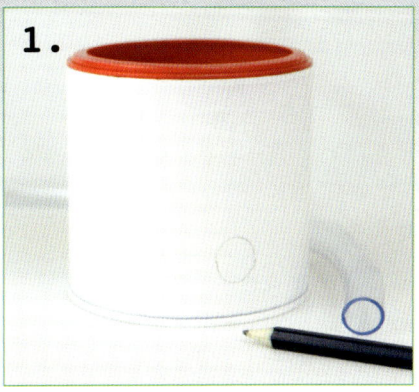

Step 1.
Use tape to attach the funnel to one end of the tube. Hold the other end of the tube against the container near the base, and draw around it with the pencil.

2.

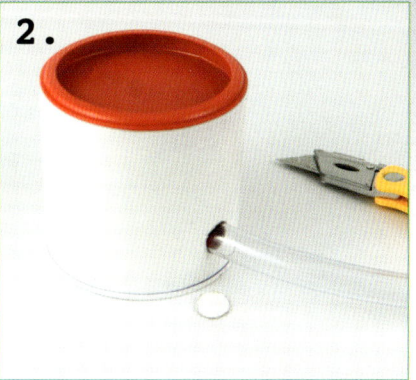

Step 2.
Ask an adult to carefully cut out the circle using the craft knife. Check the tube fits into the hole with a little space to spare (this will let air out when you sing into it).

3.

Step 3.
Take the lid off the container and run the pencil around on top of it, just inside the rim, to draw a circle on the lid. Ask an adult to use a craft knife to cut out this circle, too.

4.

Step 4.
Blow up the balloon to stretch it, let it down, then lay it flat. Use the scissors to cut the open end off, just below the neck. Stretch the balloon over the open top of the container and pull it tight.

5.

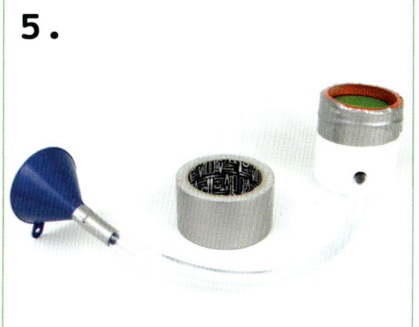

Step 5.
Now take the lid with the hole cut in it, and push it firmly back onto the container to hold the balloon in place. Wrap tape around the edge to fix the lid, balloon and container together.

6.

Step 6.
Sprinkle a thin layer of glitter onto the flat balloon surface. Fit the tube into the hole in the side of the container, and put the funnel end to your mouth.

7.

Step 7.
Now gently sing a long, single note into the eidophone funnel. Higher notes work best. Each note should make the glitter on the surface form a different pattern.

WAVE VIBRATIONS

The eidophone's amazing patterns happen because of the way thin, flat surfaces vibrate. Usually, the whole surface doesn't just shake to and fro at once. Instead, some parts move up while others move down, setting up waves known as standing waves. The areas in between the moving parts stay still. So if there is liquid or powder on the surface, it will be pushed towards these stiller areas, revealing the wave patterns.

Megan Watts Hughes created different patterns by using varying pitches.

GET INTO THE GROOVE

This real, working record player shows how sounds are stored in a record's grooves.

MAKER PROFILE:

Thomas Edison
(1847–1931)

US inventor Thomas Edison worked on many things, including light bulbs, film cameras and sound recording. His 1877 creation, the phonograph, was the first machine that could record and play back sound. It worked by making sound vibrations move a needle, which scratched a pattern in a groove on a moving record.

At first, sounds were recorded on cylinder-shaped records. These were later replaced by flat vinyl discs. Although we now have newer recording methods, vinyl records are still popular, as they have a unique sound quality.

A photograph of Edison with an early phonograph, taken in 1878.

WHAT YOU NEED

- a piece of A3 paper
- a thin, sharp needle or pin
- sticky tape
- a piece of A4 card
- scissors
- enough felt to cover two books
- paperclips or bulldog clips
- two paperback books, the same thickness
- pencil
- an old vinyl record that you don't mind getting scratched

If you don't have a vinyl record to use, you can get one cheaply at a charity shop.

20

1.

Step 1.
Take the sheet of A3 paper and curl it around to make a cone shape that's closed at the narrow end. Fix it in place with sticky tape.

2.

Step 2.
Push the pin or needle through the cone, about 2 cm from the pointed end, so that it sticks out at a right angle.

3.

Step 3.
Fold the piece of card in half and cut a large semicircle shape in the folded side. Fold up the two ends of the card to make a stand for the cone to rest in.

4.

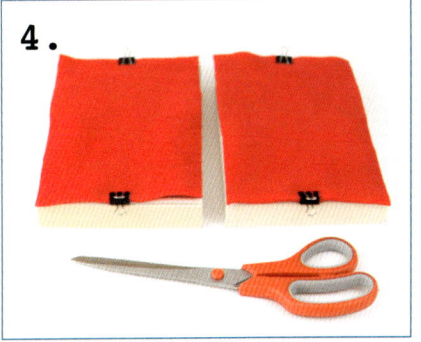

Step 4.
Put the two books flat on a table with a space between them. Cut two pieces of felt to fit on top of the books, and use clips to hold them in place.

5.

Step 5.
Push the pencil through the middle of the record, and stand it in between the books, so that the pencil rests on the table and the record rests on the felt. This will help you keep the record flat as you turn it.

6.

Step 6.
Position the cone and its stand next to the record, and carefully place the tip of the pin or needle on the record, angled slightly sideways, like this.

7.

Step 7.
Now rotate the record clockwise as steadily as you can, using both hands to turn the pencil. It may take practice, but once you have a steady speed, you should be able to hear the record playing through the cone.

STORED VIBRATIONS

The record has a long spiral groove running around and around it. Sounds are recorded into the groove by using the sound waves to make a needle vibrate. It scratches bumps and patterns into the groove. When you run a needle through it again, the bumps make the needle vibrate and make the same sound vibrations. The vibrations pass into the cone, which channels and directs them, making them easier to hear.

SONIC
SCULPTURE

Create your own noise-making sound art installation, based on a marble run.

MAKER PROFILE:

Kym Maxwell
(1972–)
Dirk Leuschner
(1962–)

Kym Maxwell is an Australian artist and teacher who often works with children to make interactive art installations. Working with furniture maker Dirk Leuschner, she created a marble run designed to make a series of different sounds as the marbles move through it. This is an example of sound art or sonic art – art that makes noises for the audience to listen to. The sounds can be made by making a sculpture with parts that move and vibrate, or by playing back recordings of any kind of sound.

Kym Maxwell and Dirk Leuschner's work *Marble Run 2: The Materials are Listening*, 2013.

WHAT YOU NEED

- a large corrugated cardboard box
- sticky tape
- scissors
- marbles
- a selection of objects to use to build the marble run and make sounds, such as: card, funnel, plastic and metal tubes, paper cups, springs, nails, wooden skewers, bells, balloons, paper clips, foil plates, old CDs, pliers, cutlery, metal wire, coins, buttons, metal washers, xylophone

1.

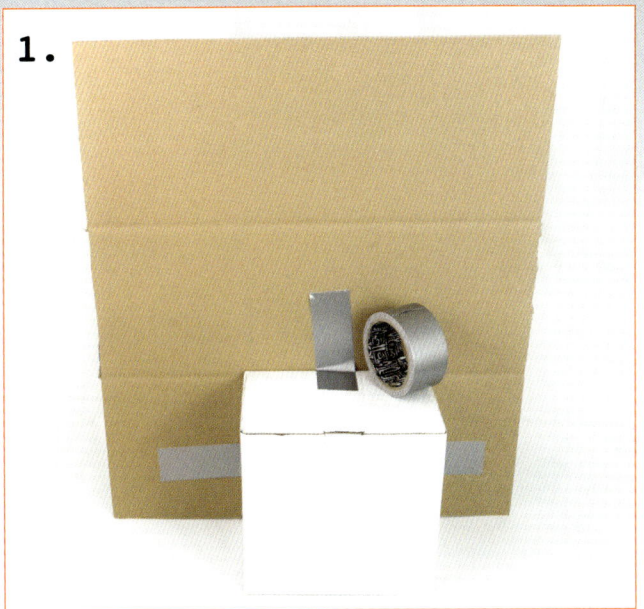

2.

Step 1.

Cut the largest side off your cardboard box to use as a base for your marble run. Lean it up against a wall or table to make a sloping surface, or tape it to another box as a support.

Step 2.

You can now start to attach sections and stages to the box for the marbles to run down. To fit as many in as possible, it helps to make each stage run across the box, from one side to the other.

> You can experiment with different ways of channelling the marbles and making sounds. The following steps are just some ideas - you can come up with your own, too.

3.

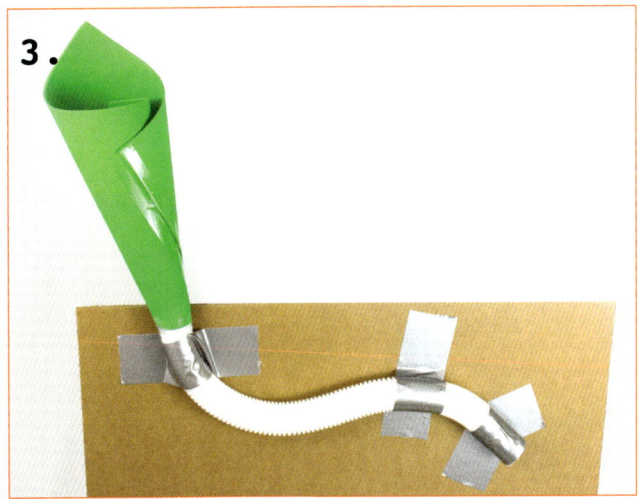

4.

Step 3.

Make a funnel from card to direct the marbles into a corrugated tube. Use strong tape to attach the tube to the cardboard. Make some parts of the tube flat so the marbles run against it to make a sound.

Step 4.

Cut the neck off a balloon and stretch the large section over a funnel to create a drum for the marbles to bounce off. Curve thick card around the drum to catch and redirect the marbles.

Step 5.
Push a row of toothpicks or skewers into the cardboard base to make a bouncy, rattly pathway.

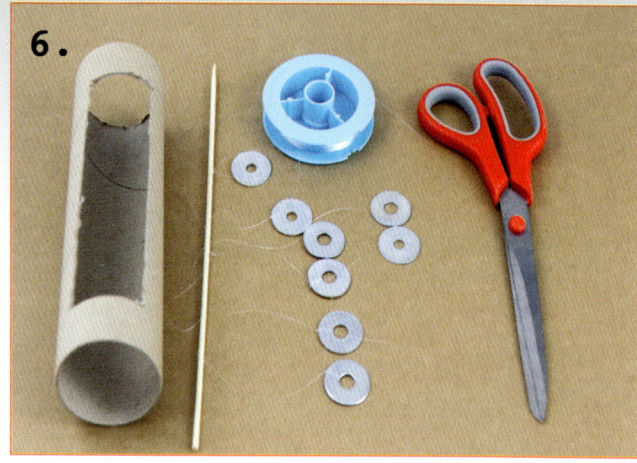

Step 6.
Cut a cardboard tube open and use thread to hang small metal objects along it, for the marbles to hit.

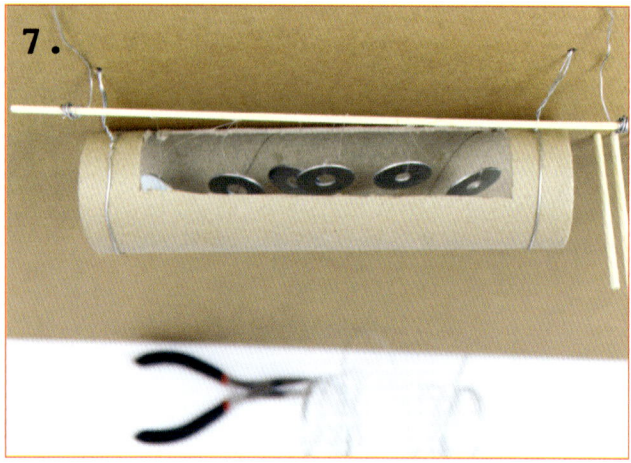

Step 7.
Use metal gardening wire and pliers to attach the cardboard roll to the base.

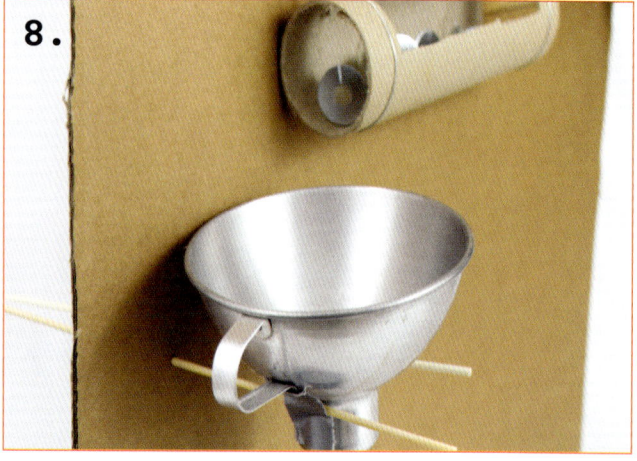

Step 8.
A large metal funnel with a wide spout will make a loud clang when the marbles hit it.

Step 9.
You can build musical instruments into your sonic marble run too, such as a small xylophone.

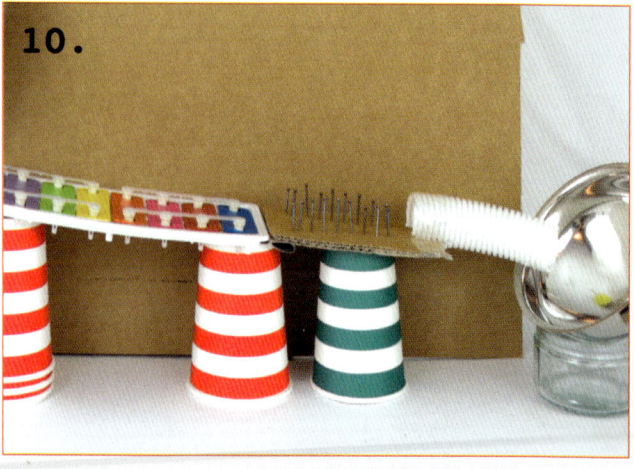

Step 10.
Create a narrow passage using nails so that the marbles hit them as they go past.

11.

Step 11.

Finish your run with a collection area for the marbles. Try different materials to get a variety of sounds.

13.

Step 13.

When you're happy with your sonic marble run, you could make a video of it working from start to finish.

Step 12.

Keep a pile of marbles handy, and test each step as you make it to see if it works. When you've made all the steps, test the whole marble run. Change or add bits to fix any problems and make sure it all runs smoothly.

MILLIONS OF MARBLES

Several artists have experimented with marble runs as a form of art. Though a marble run can have lots of complicated stages, it's a very simple concept and doesn't need a power supply, as the structure uses gravity to make the marble travel along the whole circuit. However, some do have a powered system that takes the marbles from the bottom and feeds them back in at the top.

ART WITH SOUNDS

Sound art isn't a very recent invention – it dates back over a hundred years. Luigi Russolo (1885–1947), an Italian painter and musician, is usually said to be the first sound artist. He built machines designed to recreate the noises of everyday life, machines and factories, which he called the 'Intonarumori'. Many artists since then have made sonic art, but it has become especially popular in the 21st century. Digital sound processing and storage, and the ability to harvest all kinds of sounds from the Internet, have given sound artists many new ways to work.

INTRUDER ALERT!

Is someone trying to sneak into your room when you're not there? BZZZZZ! Catch them with this DIY electric pressure mat alarm.

MAKER PROFILE:

Augustus Russell Pope
(1819–1858)

Augustus Russell Pope was a US church minister and part-time inventor. In 1853, he patented his invention for an electric alarm system triggered by someone entering a door or window. This would complete an electrical circuit and make a bell ring. (Until then, people just used guard dogs, or basic alarms such as a hanging bell attached to the door.) Pope himself died not long afterwards, but his ideas were used to make the first commercial electric burglar alarms.

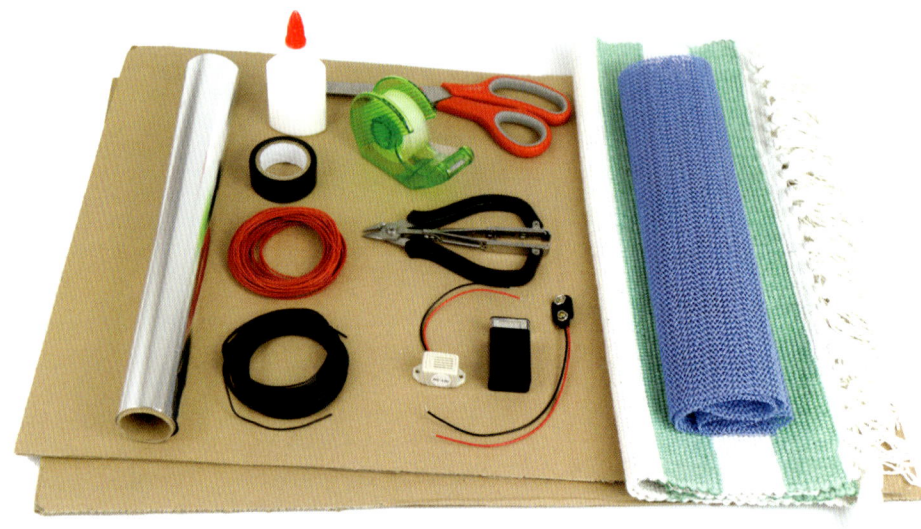

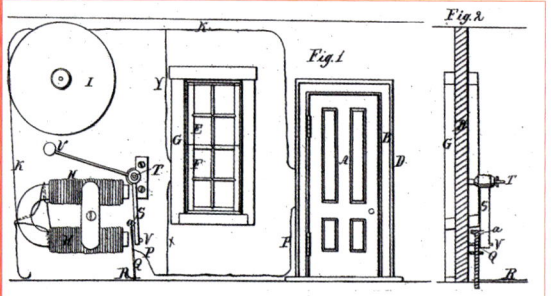

A diagram by Augustus Russell Pope, showing how his invention worked.

To be applied to either a door or window, or both, of a dwelling-house or other building, for the purpose of giving alarm in case of burglarious or other attempts to enter through said door or window.
– *Augustus Russell Pope*

WHAT YOU NEED

- two large rectangles of corrugated cardboard
- aluminium foil
- scissors
- sticky tape
- insulated electrical wire
- wire cutters and strippers
- electrical tape
- a 9V battery
- a battery connector
- a small electric buzzer
- felt, thin foam or rug underlay
- glue
- a door mat or rug

1.

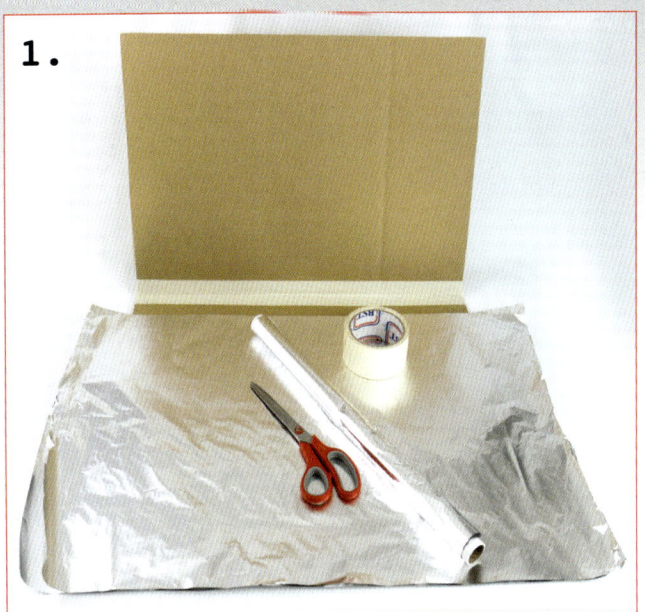

Step 1.

Join the pieces of cardboard together with tape to make the folding pressure mat. Cut two pieces of aluminium foil slightly larger than the folded cardboard. Open the cardboard out and put one piece of foil on each side.

2.

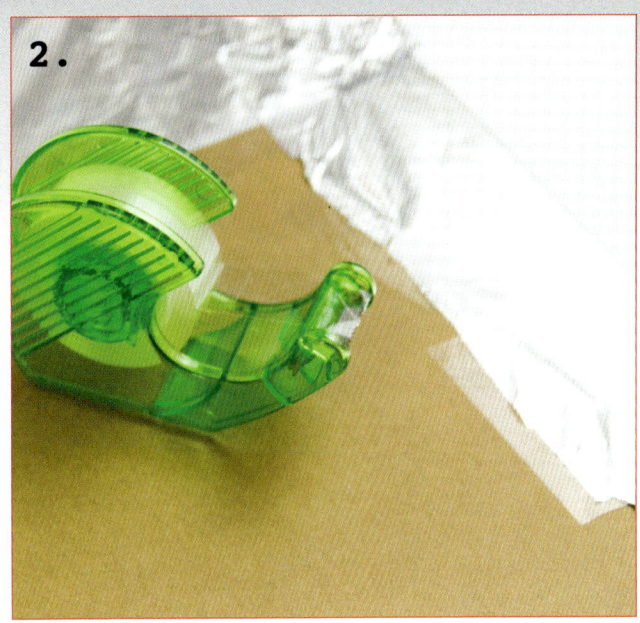

Step 2.

Fold the foil over the outer edges of the cardboard, and stick it in place using tape. In the middle of the cardboard, stick the edges of the foil down with tape, making sure the two pieces of foil do not touch.

3.

Step 3.

Cut two pieces of electrical wire, each about 30 cm long. Use the wire strippers to strip about 1 cm of the plastic coating off both ends of each wire.

4.

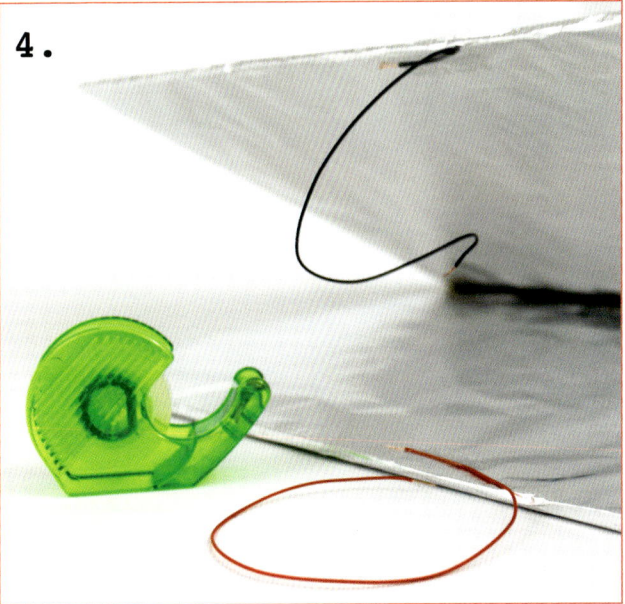

Step 4.

Open out the cardboard and push one of the wire ends into the foil, near the outer edge. Fix it in place with sticky tape, making sure the bare wire is touching the foil. Do the same with the other wire on the other side.

5.

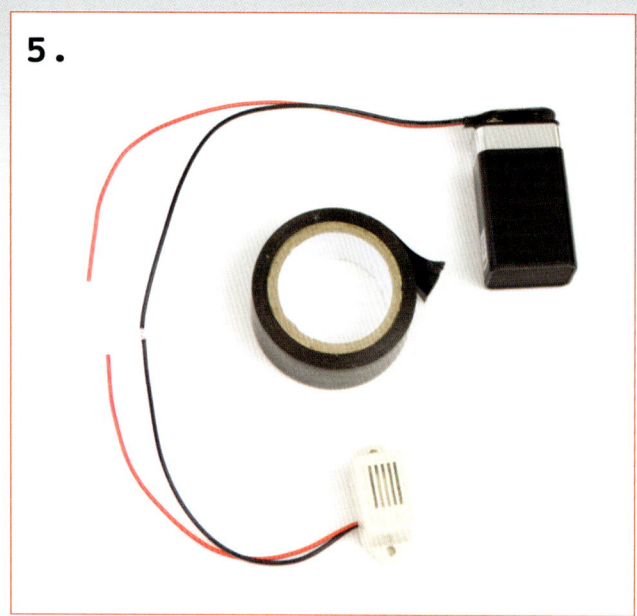

Step 5.

Press the connector onto the battery. Attach the black wire on the connector to the black wire on the buzzer, by twisting the bare ends together, then wrapping electrical tape firmly around them.

6.

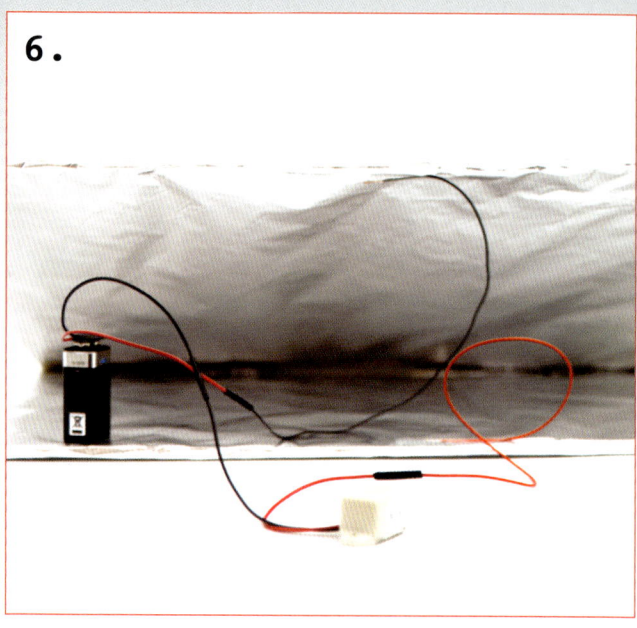

Step 6.

In the same way, attach the free wire on the connector to the wire connected to one side of the mat. Now attach the free wire on the buzzer to the wire connected to the other side of the mat.

7.

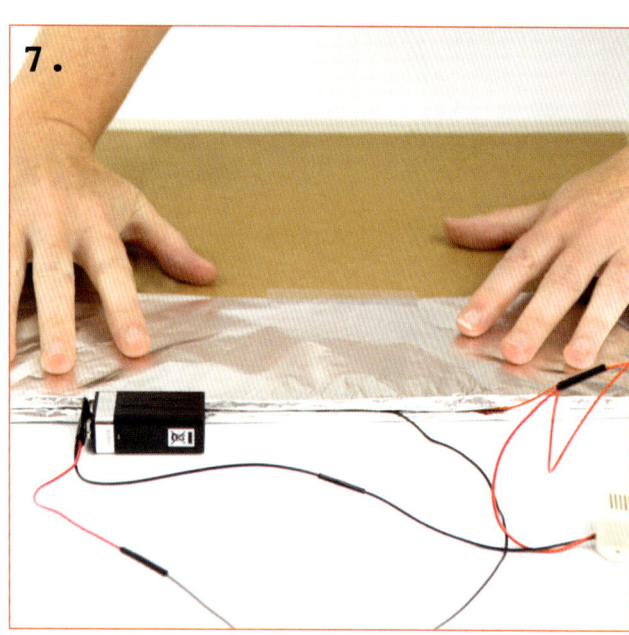

Step 7.

Now test your circuit by pressing the two layers of foil together. This should complete the circuit and make the buzzer sound. (If it doesn't, check your battery has power and all your connections are well-attached.)

8.

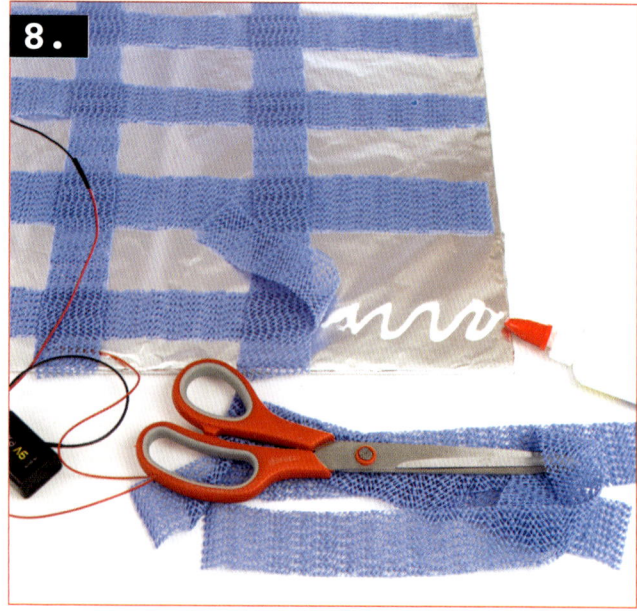

Step 8.

Cut several strips of felt, thin foam or rug underlay, about 2 cm wide. Open out the pressure mat and lay the strips along and across one side in a grid pattern. Use a little glue to stick the strips in place, let it dry, then trim the strips to the right length around the edges. This will help keep the two layers of foil apart, until someone steps on the mat.

9.

Step 9.

Close the mat and check the foil does not touch and the buzzer does not sound. Hide the mat under a rug, just outside or inside a doorway, with the wires, battery and buzzer tucked out of sight. Now wait for someone to stand on the mat and listen to the buzzer go off!

CLOSING THE CIRCUIT

Like Augustus Russell Pope's invention, this alarm works by completing an electric circuit when the 'burglar' tries to enter. Electric circuits can only work once all the parts of the circuit are connected together, and electricity from the battery can flow through them. Standing on the mat connects the two pieces of foil, completing the circuit.

As the electric current flows through the buzzer, it travels around a coil of wire surrounding a metal bar. The current turns the bar into a magnet, and it is pulled towards a metal surface, making a sound. But this also breaks the circuit inside the buzzer, switching it off. The bar moves back, connecting the circuit again, repeating the cycle. The bar repeatedly moves to and fro, hitting the metal at high speed, which makes a buzzing sound.

Electric pressure mats can also be used in automatic door systems. When someone steps on the mat, they complete a circuit, and the door in front of them opens!

When electricity runs through the coil, the rod in the middle becomes magnetic, and it moves down.

When the middle rod moves down, this breaks the circuit, and it moves back.

The repeated movement makes a buzzing sound.

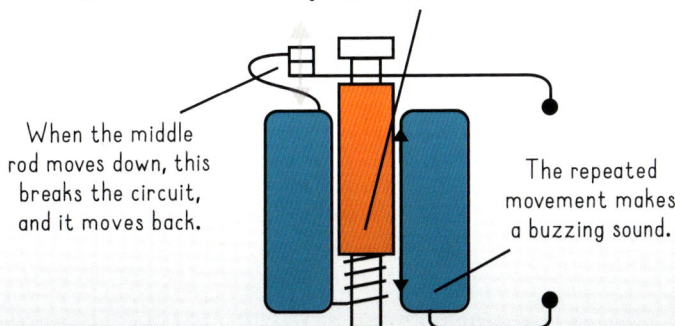

GLOSSARY

amplifier A device used to increase the power of a sound signal.

buzzer An electric device that makes a buzzing sound when electricity flows through it.

digital sound processing Converting sounds into computer data, and altering them in this form to change what they sound like.

distortion Deliberate or accidental changes made to sounds as they pass through electric devices.

eidophone A device for using voice sound waves to make a membrane vibrate, revealing different patterns for different notes.

electric circuit A loop of wires and components that an electric current can flow around.

electric current A flow of electricity.

energy The power to make things happen, move or change.

frequency The speed of a sound vibration.

hydrophone A device for listening to sounds made underwater.

installation A type of 3D artwork that is built or installed into a particular space.

interactive art Art that involves the viewer or audience, using them to make things happen or to become part of the artwork.

Intonarumori A collection of experimental sound art instruments made by Italian artist Luigi Russolo in the early 1900s.

membrane A thin, flexible sheet of material, such as the eardrum or a drum skin.

microphone A device for collecting sound waves and converting them into electrical signals.

molecule A tiny single unit of a substance, made from even smaller units called atoms.

percussion Musical instruments that make sounds by being struck or shaken.

phonograph The name given to the first sound recording and playback machine, invented by Thomas Edison.

pitch How high or low a sound is.

sonic To do with sound.

sound waves Waves made up of molecules vibrating to and fro, which spread out through the air or another substance as sound travels through it.

speaker A device that converts an electrical signal into a sound that can be heard.

standing wave A vibration of a surface or material that makes some parts move to and fro while other parts stay in one place.

vibrate To move quickly and regularly to and fro.

vinyl record A disc-shaped piece of vinyl (a type of plastic) used to record sounds on.

volume The loudness of a sound, which is related to the amount of energy in the vibrations it makes.

FURTHER INFORMATION

WEBSITES ABOUT SOUND

Exploratorium Science Snacks: Sound
www.exploratorium.edu/snacks/subject/sound

BBC Radiophonic Workshop: Tracks
www.bbc.co.uk/music/artists/39f0d457-
37ba-43b9-b0a9-05214bae5d97

Science Kids: Sound for Kids
www.sciencekids.co.nz/sound.html

Soundation: Free online audio editor
https://soundation.com/

WEBSITES ABOUT MAKING

Tate Kids: Make
www.tate.org.uk/kids/make

PBS Design Squad Global
https://pbskids.org/designsquad

Instructables
www.instructables.com

Make:
https://makezine.com/

WHERE TO BUY MATERIALS

Maplin
For electronic components and making projects
www.maplin.co.uk

Hobbycraft
For art and craft materials
www.hobbycraft.co.uk

B&Q
For pipes, tubing, wood, glue and other hardware
www.diy.com

Fred Aldous
For art and craft materials, photography
supplies and books
www.fredaldous.co.uk

BOOKS

Science in a Flash: Sound
by Georgia Amson-Bradshaw (Wayland, 2017)

Home Lab by Robert Winston and Jack
Challoner (Dorling Kindersley, 2016)

Tabletop Scientist: The Science of Sound
by Steve Parker (Dover Publications, 2013)

Project Code: Create Music with Scratch
by Kevin Wood (Franklin Watts, 2017)

*Junkyard Jam Band: DIY Musical
Instruments and Noisemakers*
by David Erik Nelson (No Starch Press, 2015)

Sound (Moving up with Science)
by Peter Riley (Franklin Watts, 2016)

Science in Infographics: Light and Sound
by Jon Richards (Wayland, 2017)

PLACES TO VISIT

**National Science and Media Museum,
Bradford, UK**
www.scienceandmediamuseum.org.uk

Science Museum, London, UK
www.sciencemuseum.org.uk

**Musical Instrument Museum, Phoenix,
Texas, USA**
https://mim.org/

INDEX

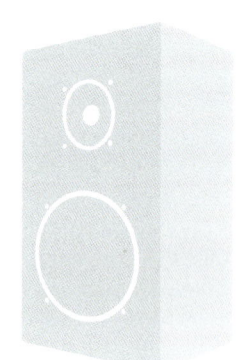